Panthalassa: The Superocean

The Vast Body that Shaped the Ancient Earth

Authored by

Zahid Ameer

Published by

Goodword eBooks

DEDICATION

"I dedicate this book to my beloved parents, whose wisdom I hold in the highest regard. Their every word of guidance has been a beacon of light, illuminating the path of my life and shaping the essence of who I am."

The Panthalassa

Contents:

The Panthalassa

Chapter 1: Introduction to Panthalassa

Panthalassa, the ancient "Superocean," was a vast and all-encompassing body of water that surrounded the supercontinent Pangaea. During much of Earth's ancient history—particularly in the late Paleozoic and early Mesozoic eras—Panthalassa dominated the planet, playing a key role in shaping its climate, geography, and the development of life. Understanding Panthalassa not only provides insight into the past but also helps us comprehend the forces that shaped the Earth's present-day landscapes and oceans. In this chapter, we will delve into what Panthalassa was, how it formed, and its immense influence on the early Earth.

1.1 What is Panthalassa?

Panthalassa (from the Greek words "pan" meaning "all" and "thalassa" meaning "sea") was a superocean that surrounded Pangaea, the single massive landmass that once dominated the Earth's surface. Existing approximately 300 million years ago, Panthalassa was the largest ocean ever to exist on Earth, covering around 70% of the planet's surface. Its scale was nearly unfathomable, stretching from the North Pole to the South Pole and spanning much of the planet's circumference.

This ocean was vast and uninterrupted, unlike the oceans of today, which are divided by landmasses. The absence of large land barriers allowed Panthalassa to host massive ocean currents and complex climate systems. It was, in many ways, the beating heart of the planet's climate, influencing temperatures, precipitation patterns, and weather systems across Pangaea. Panthalassa was not only significant in terms of its size but also its role as a mediator of Earth's climate and its capacity to sustain life in its vast marine ecosystems.

1.2 The Formation of Panthalassa

The formation of Panthalassa is directly linked to the geological and tectonic events that shaped the Earth during the late Paleozoic era, roughly 300 million years ago. During this time, Earth's tectonic plates were slowly but steadily shifting, driven by forces deep within the Earth's mantle. These plates carried the Earth's landmasses, which over hundreds of millions of years had been slowly drifting apart and then converging. Eventually, nearly all of Earth's continents collided and merged into one supercontinent known as **Pangaea**.

The creation of Pangaea did not happen overnight. It was the culmination of complex tectonic processes that began during the earlier Paleozoic era. As tectonic plates shifted, ancient oceans like the Iapetus Ocean and Rheic Ocean disappeared as the continental masses closed in on each other. This collision of landmasses gradually formed the supercontinent Pangaea around 335 million years ago. With the continents united into a single landmass, the surrounding waters coalesced into a single massive ocean—Panthalassa.

While Panthalassa stretched across most of the planet, other bodies of water existed as well. One notable feature within Pangaea was the **Tethys Sea**, a smaller inland sea that lay between the northern and southern parts of Pangaea. However, Panthalassa overshadowed these smaller seas in terms of size, depth, and influence.

The formation of Panthalassa and Pangaea was not simply a matter of continents joining together; it was also driven by complex tectonic dynamics such as **subduction** and **mid-ocean ridges**. Subduction, where one tectonic plate moves under another and sinks into the mantle, played a key role in shaping Panthalassa's ocean floor. At the same time, mid-ocean ridges, where new oceanic crust is created through volcanic activity, helped to expand and reshape the seafloor of Panthalassa over time. The ocean floor was continually recycled through these processes, making Panthalassa a geologically active environment.

1.2.1 The Role of Plate Tectonics

The tectonic activities that formed Panthalassa were part of Earth's ongoing **supercontinent cycle**, where landmasses periodically come together and break apart. This cycle, spanning hundreds of millions of years, is driven by the motion of Earth's tectonic plates. As continents collided, creating mountain ranges and pushing up landmasses, vast subduction zones formed around the perimeter of Panthalassa. These zones were regions where the oceanic crust of Panthalassa was being recycled back into the Earth's mantle, triggering volcanic activity and forming deep ocean trenches.

One of the most significant geological features of Panthalassa was the **Panthalassic Ridge**, a mid-ocean ridge system where new oceanic crust was continually being formed. This ridge was similar to the modern **Mid-Atlantic Ridge**, though it spanned a much greater distance. The ridge system acted as a conveyor belt, driving the oceanic plates outward and contributing to the dynamic nature of the superocean's floor.

1.2.2 The Influence of Panthalassa on Earth's Geology

Panthalassa's formation also had a profound effect on Earth's geological features. The presence of such a vast ocean influenced the distribution of heat across the globe, creating climate gradients that were distinct from today's climatic patterns. Coastal regions of Pangaea, where land met Panthalassa, experienced a humid and temperate climate due to the proximity to the ocean, while the interior of Pangaea, far from the moderating influence of the ocean, was arid and desert-like. This contrast between coastal and inland climates was a direct result of Panthalassa's immense size and its ability to store and distribute heat.

The massive expanse of Panthalassa also contributed to **oceanic circulation** patterns that were likely more intense than anything seen in the modern world. The lack of land barriers meant that ocean currents could circulate freely around the globe, creating powerful gyres and influencing wind patterns. These oceanic currents played a key role in distributing nutrients and supporting life in the marine ecosystems of Panthalassa.

1.3 Panthalassa's Geological Legacy

Although Panthalassa no longer exists in its ancient form, its legacy lives on in the modern Pacific Ocean. When Pangaea began to break apart during the Mesozoic era, around 175 million years ago, Panthalassa began to fragment as well. The breakup of Pangaea marked the end of the supercontinent era and the beginning of the formation of the Atlantic, Indian, and Southern Oceans. However, the largest remnant of Panthalassa is today's **Pacific Ocean**, which covers a vast portion of Earth's surface and retains many of the features of the ancient superocean.

The Pacific Ocean still exhibits deep ocean trenches, such as the **Mariana Trench**, and extensive subduction zones along the **Ring of Fire**, which are remnants of the tectonic processes that once shaped Panthalassa. The ocean's tectonic activity continues to influence volcanic eruptions, earthquakes, and the creation of new seafloor, just as it did during the reign of Panthalassa.

As we continue to explore and study Earth's geological history, the story of Panthalassa provides critical insight into how oceans shape the planet's surface, climate, and life. This superocean was not only a dominant feature of Earth's ancient past but also a precursor to the modern oceans that regulate our planet's climate and ecosystems today.

Chapter 2: Geological Evolution of Panthalassa

The vast ocean of Panthalassa, which enveloped the supercontinent Pangaea, was not a static or unchanging body of water. Rather, it was a dynamic, ever-evolving ocean shaped by powerful geological forces. This chapter explores how tectonic activity, continental drift, and the breakup of Pangaea all played a critical role in the transformation of Panthalassa into modern-day oceans. These processes not only reshaped the Earth's surface but also had profound effects on climate, sea levels, and marine ecosystems over millions of years.

2.1 Tectonic Activity and Panthalassa

Panthalassa, like all large bodies of water on Earth, was underpinned by tectonic processes. The ocean's floor was constantly being reshaped by the movement of the Earth's lithospheric plates. The superocean, which spanned almost the entire globe, was bordered by mid-ocean ridges, deep-sea trenches, and numerous volcanic arcs, all of which contributed to the dynamic nature of Panthalassa's geology.

Mid-Ocean Ridges and Seafloor Spreading

The formation of Panthalassa's oceanic crust occurred primarily at mid-ocean ridges, long, volcanic mountain chains found beneath the surface of the water. These ridges are key sites of seafloor spreading, where new oceanic crust is continuously generated as magma from the Earth's mantle rises and solidifies at the ridge axis. This process pushed older crust away from the ridge, causing the oceanic plates to move apart.

In Panthalassa, the mid-ocean ridges were immense, stretching for thousands of kilometers. The seafloor spreading process was critical in the expansion and reshaping of Panthalassa's oceanic crust. As new crust formed, it drove tectonic plates across the surface of the ocean. The plates' movement caused them to interact with the continental landmasses of Pangaea, influencing coastal uplift, volcanic activity, and the creation of island arcs, which in turn affected the ecosystems of both the ocean and the land.

Subduction Zones and Oceanic Trenches

While new oceanic crust was continuously forming at mid-ocean ridges, older crust was being consumed at subduction zones—regions where one tectonic plate was forced beneath another and reabsorbed into the mantle. Subduction zones, often found at the edges of Panthalassa, formed deep oceanic trenches, some of which reached extreme depths (similar to the Mariana Trench today). These trenches were some of the most geologically active areas on the planet, hosting intense seismic and volcanic activity.

Subduction in Panthalassa created chains of volcanic islands and triggered powerful earthquakes and tsunamis. The collision and subduction of oceanic plates generated vast volcanic arcs, which formed island chains along the boundaries of the ocean, such as the ancient equivalents of the modern-day Pacific "Ring of Fire." These regions were not only tectonically active but also served as hotspots for biodiversity, with unique ecosystems developing around volcanic vents in deep ocean waters.

Earthquakes and Volcanic Activity

Tectonic activity in Panthalassa's subduction zones resulted in frequent volcanic eruptions and powerful earthquakes. The intense pressure from subducting plates caused large portions of the ocean floor to buckle and fold, leading to major seismic events. These earthquakes often triggered undersea landslides and tsunamis, which could have significant effects on coastal environments around Pangaea.

Volcanism in Panthalassa contributed to the buildup of large volcanic islands and mountain chains, both on the ocean floor and at the edges of the continental landmasses.

Volcanic eruptions released gases and ash into the atmosphere, which could have influenced the global climate, while the underwater eruptions created hydrothermal vents that supported unique deep-sea ecosystems.

2.2 Panthalassa and the Supercontinent Cycle

Earth has undergone several supercontinent cycles throughout its history. These cycles are characterized by the assembly and breakup of massive landmasses, accompanied by the formation and destruction of superoceans. Panthalassa played a central role in one such cycle during the late Paleozoic and early Mesozoic eras, when it surrounded the supercontinent Pangaea.

The Assembly of Pangaea and Panthalassa

The formation of Pangaea, which began around 335 million years ago in the late Carboniferous period, was a slow and complex process. It involved the collision of several smaller continents and landmasses, such as Laurentia, Gondwana, and Siberia. As these landmasses merged, they formed a single supercontinent that was surrounded by a vast ocean—Panthalassa.

Panthalassa essentially became the single, dominant ocean on Earth. Smaller seas, such as the Tethys Sea, existed between certain portions of Pangaea, but Panthalassa covered most of the planet's surface. The ocean played a crucial role in regulating the climate of the time. Its immense size allowed for the circulation of ocean currents that distributed heat across the planet, helping to maintain relatively stable temperatures despite the shifting positions of the continents.

The Breakup of Pangaea and the Fragmentation of Panthalassa

The supercontinent cycle continued into the Mesozoic era, when Pangaea began to break apart. Around 175 million years ago, during the Jurassic period, tectonic forces caused the supercontinent to fracture into smaller land masses. This breakup initiated the fragmentation of Panthalassa as well.

The process of Pangaea's breakup occurred in several stages. First, the northern and southern halves of the supercontinent—Laurasia and Gondwana—began to drift apart. This rift marked the beginning of the formation of the Atlantic Ocean. As the landmasses continued to move, new oceans formed in the spaces between the drifting continents, and Panthalassa gradually shrank.

By the time the Cretaceous period arrived, Panthalassa had significantly diminished in size. The Tethys Sea expanded between Gondwana and Laurasia, and the formation of the Indian Ocean began. Panthalassa's western regions, once

The Panthalassa
bordered by Pangaea, slowly transformed into what would become the Pacific Ocean. This process of continental drift and oceanic transformation continued into the Cenozoic era, giving rise to the modern configuration of Earth's continents and oceans.

2.3 The Transition to Modern Oceans

The transition from Panthalassa to the modern oceans was a long and gradual process that took place over tens of millions of years. As Pangaea continued to fragment, new ocean basins formed, and Panthalassa itself evolved into the Pacific Ocean, while other oceans, such as the Atlantic, Indian, and Southern Oceans, began to take shape.

The Formation of the Atlantic and Indian Oceans

The breakup of Pangaea was accompanied by the creation of new rifts between the drifting continents. These rifts filled with water, leading to the formation of new ocean basins. One of the most significant developments was the formation of the Atlantic Ocean, which began as a narrow seaway between the separating landmasses of Africa and North America. Over time, the Atlantic continued to widen as the continents drifted farther apart.

Similarly, the Indian Ocean began to form as India separated from Africa and Antarctica, moving northward toward the Asian landmass. This process of continental drift and ocean basin formation contributed to the fragmentation of Panthalassa and the creation of distinct oceanic regions that we recognize today.

The Pacific Ocean: The Last Vestige of Panthalassa

As the Atlantic, Indian, and Southern Oceans grew larger, Panthalassa continued to shrink. The remaining portion of Panthalassa became the Pacific Ocean, which is now the largest ocean on Earth. The Pacific retains many of the features of its ancient predecessor, including vast mid-ocean ridges, deep-sea trenches, and extensive volcanic activity along its margins.

The Pacific Ocean is often referred to as the "last vestige" of Panthalassa because it is the only modern ocean that directly descends from the ancient superocean. The tectonic processes that shaped Panthalassa, such as seafloor spreading and subduction, continue to influence the Pacific today. The ocean's Ring of Fire, a zone of intense volcanic and seismic activity, is a remnant of Panthalassa's tectonic legacy.

Climate and Environmental Changes During the Transition

The fragmentation of Panthalassa and the formation of modern oceans had profound effects on the Earth's climate and environment. As the continents drifted apart, ocean

The Panthalassa currents changed, altering global heat distribution. The breakup of Pangaea also led to the creation of new coastlines and shallow seas, which supported diverse marine ecosystems and contributed to the evolution of new species.

The separation of landmasses also influenced atmospheric circulation patterns, leading to changes in rainfall, temperature, and wind patterns. These shifts in climate played a crucial role in shaping the evolutionary trajectories of both terrestrial and marine life during the Mesozoic and Cenozoic eras.

Chapter 3: Climate and Environment of Panthalassa

Panthalassa, the vast superocean that surrounded the supercontinent Pangaea, had a profound impact on the Earth's climate and environment during the late Paleozoic and early Mesozoic eras. This chapter delves into how the unique characteristics of Panthalassa influenced global climate patterns, the role it played in regulating glacial cycles, and its involvement in the Permian-Triassic extinction event, often referred to as "The Great Dying." Understanding Panthalassa's influence on climate helps illuminate how ancient oceans shaped the world we know today.

3.1 Climate Dynamics

The climate during Panthalassa's existence was vastly different from today's environment, largely because the Earth's landmass was concentrated into the supercontinent Pangaea, and the rest of the planet was covered by the massive expanse of Panthalassa. This unique configuration affected global temperatures, weather patterns, ocean circulation, and ultimately the distribution of life on Earth.

3.1.1 The Role of Ocean Size in Climate Regulation

The sheer size of Panthalassa meant that it acted as a gigantic heat reservoir. Water has a higher specific heat capacity than land, meaning that it can absorb and store large amounts of heat with relatively small temperature changes. As a result, Panthalassa's vast waters contributed to a more stable and uniform climate across the planet compared to today, when the Earth's surface is divided between multiple oceans and landmasses.

During Panthalassa's reign, temperature extremes between the equator and poles were less pronounced, and this relatively uniform climate had a global impact. The extensive oceanic surface of Panthalassa absorbed solar heat in the tropical regions and redistributed it towards the poles, reducing the sharp temperature gradients that we see today. This helped create a more temperate global climate, especially near coastal regions of Pangaea.

3.1.2 Ocean Currents and Heat Redistribution

Ocean currents within Panthalassa played a significant role in regulating Earth's climate. Like the modern-day ocean currents (such as the Gulf Stream and the Antarctic Circumpolar Current), Panthalassa had powerful currents that transported heat around the planet. The flow of warm and cold water masses helped maintain temperature balance and influenced weather patterns over both land and sea.

In Panthalassa, the deep ocean currents were driven by global wind patterns, as well as differences in water temperature and salinity (thermohaline circulation). Warm surface currents transported heat away from the equator, while cold, dense water sank near the poles and flowed back towards the equator. This circulation system helped moderate the climate of Pangaea, preventing it from experiencing extreme heat in the tropics and intense cold in the polar regions.

3.1.3 The Impact of Panthalassa on Weather Patterns

The interaction between Panthalassa's oceania systems and the atmosphere created unique weather patterns. The superocean's large expanse of water provided moisture for the atmosphere, fueling precipitation over the continents, particularly along coastal areas of Pangaea. However, the interior regions of Pangaea were likely much drier, as they were far removed from the ocean's moisture sources. This resulted in extreme desert-like conditions in the continental interior, a phenomenon known as "continentality."

Moreover, the prevailing wind patterns during the existence of Panthalassa were likely different from today. The planet's atmosphere at the time was affected by the placement of Pangaea along the equator, which shaped the global wind belts and storm tracks. Seasonal monsoons may have been intense along the coastal regions of Pangaea due to the interaction of warm moist air from Panthalassa and the cooler, drier air over the continent.

3.2 Glacial Cycles and Panthalassa

Panthalassa played an important role in Earth's glacial cycles, particularly during the late Paleozoic era, when the planet experienced significant ice ages. The Carboniferous and Permian periods, in particular, were marked by extensive glaciations, with massive ice sheets covering much of the southern supercontinent Gondwana (part of Pangaea).

3.2.1 Carboniferous Ice Age and Gondwanan Glaciation

The Carboniferous period (approximately 360 to 300 million years ago) was a time of significant glaciation, especially in the southern hemisphere. The supercontinent Gondwana, which was located near the South Pole, was covered by extensive ice sheets, and Panthalassa's waters played a critical role in regulating these glacial cycles.

Panthalassa's currents transported warm and cold water masses across the planet, which helped control the growth and retreat of glaciers. When the planet entered glacial periods, cold water from the polar regions would spread across Panthalassa's surface, contributing to global cooling. Conversely, during interglacial periods, warmer ocean currents would flow towards the poles, leading to glacial melting and rising sea levels.

3.2.2 Sea-Level Fluctuations and Climate Change

During glacial periods, a large amount of the planet's water was locked in ice sheets, leading to lower sea levels. Panthalassa's vast expanse would have experienced significant sea-level fluctuations in response to these ice ages. These fluctuations not only affected coastal environments and marine life but also played a role in the global climate by altering oceanic circulation patterns.

In the Permian period (approximately 298 to 252 million years ago), glacial cycles became less pronounced, but Panthalassa continued to influence Earth's climate. By regulating the exchange of heat and moisture between the ocean and atmosphere, Panthalassa helped moderate the transition from a cold, glaciated Earth to a warmer, more temperate climate as the planet moved towards the Mesozoic era.

3.3 The Great Permian Extinction

One of the most significant and devastating events in Earth's history occurred at the end of the Permian period: the Permian-Triassic extinction event, also known as "The Great Dying." This event, which took place around 252 million years ago, wiped out over 90% of marine species and 70% of terrestrial species, making it the most severe mass extinction event in Earth's history. Panthalassa's vast waters were a central stage for the dramatic environmental changes that led to this extinction.

3.3.1 Causes of the Permian-Triassic Extinction

The exact causes of the Permian-Triassic extinction event are still debated, but the most widely accepted theory involves massive volcanic eruptions in what is now Siberia, known as the Siberian Traps. These eruptions released enormous quantities of carbon dioxide and other greenhouse gases into the atmosphere, triggering a cascade of environmental catastrophes.

The sudden increase in atmospheric carbon dioxide led to global warming, causing temperatures to rise by as much as 10°C (18°F) over a short geological period. This rapid warming had devastating effects on both terrestrial and marine ecosystems. In Panthalassa, the warming caused a breakdown in oceanic circulation, leading to widespread oceanic anoxia, where oxygen levels in the deep ocean became dangerously low.

3.3.2 Ocean Acidification and Anoxia

In addition to global warming, the increased levels of carbon dioxide dissolved into Panthalassa's waters, leading to ocean acidification. Acidic waters reduced the availability of carbonate ions, which are essential for marine organisms that build shells and skeletons, such as corals, mollusks, and some plankton. The collapse of these organisms at the base of the food chain led to a cascading effect throughout marine ecosystems.

At the same time, the disruption of ocean circulation and stratification of Panthalassa's waters caused large areas of the ocean to become anoxic, meaning that oxygen levels plummeted. Without sufficient oxygen, marine life could not survive in these regions, leading to massive die-offs of marine species. The extinction was particularly severe in shallow marine environments, where many species of coral, brachiopods, and trilobites vanished entirely.

3.3.3 Impact on Panthalassa's Ecosystems

The Permian-Triassic extinction event had a devastating impact on Panthalassa's ecosystems. Coral reefs, which had flourished in the warm, shallow waters along the edges of Pangaea, were almost entirely destroyed. Large marine reptiles, fish, and invertebrates also suffered significant losses, with many species driven to extinction.

The recovery from this mass extinction was slow. It took millions of years for marine ecosystems to stabilize and for new species to evolve and repopulate the oceans. The Triassic period, which followed the extinction event, saw the emergence of new marine fauna, including early ancestors of modern fish, marine reptiles, and invertebrates, but Panthalassa's once-thriving ecosystems were forever altered.

Conclusion

The climate and environment of Panthalassa were central to shaping the Earth's history during the Paleozoic and early Mesozoic eras. From its role in regulating global temperatures and ocean currents to its impact on glacial cycles and mass extinction events, Panthalassa was a key player in the evolution of the planet's climate and life. Understanding the dynamics of this ancient superocean offers valuable insights into the complex interplay between oceans, climate, and life, both in the distant past and in our modern world.

Chapter 4: Marine Life in Panthalassa

The ancient ocean of Panthalassa, a vast and seemingly endless body of water surrounding the supercontinent Pangaea, was a dynamic ecosystem teeming with life. Though the biodiversity patterns of Panthalassa were distinct from those found in modern oceans, this superocean was home to a rich and diverse array of marine organisms. From early fish species to massive marine reptiles, and from small invertebrates to large predators, Panthalassa's ecosystem was complex and fascinating. This chapter delves into the biodiversity, marine reptiles, and shallow ecosystems that defined life in this ancient ocean.

4.1 Biodiversity in the Superocean

Panthalassa, being the largest ocean of its time, supported a wide variety of marine organisms, many of which were uniquely adapted to the ocean's environmental conditions. The biodiversity of Panthalassa was heavily influenced by its immense size, its depth, and the different climates found along its vast expanse.

Early Fish and Primitive Vertebrates

One of the most significant developments in the oceans of the Paleozoic and early Mesozoic eras was the evolution of early fish and primitive vertebrates. During the Devonian period, often called the "Age of Fishes," marine life in Panthalassa experienced a major evolutionary explosion. Jawless fish such as the agnathans were common, but more advanced species, including armored fish like placoderms, were also prevalent. These early fish paved the way for the evolution of more modern bony fish (osteichthyans) and cartilaginous fish (sharks and rays) that began to dominate marine ecosystems later on.

In Panthalassa, some fish species were adapted to deeper waters, while others thrived near the surface. Many species, especially early ray-finned fish, evolved to exploit the diverse niches available within the vast ocean. Large predatory fish were common in the open ocean, while smaller, more agile fish inhabited the shallow waters along Panthalassa's coastlines.

Invertebrates and Crustaceans

The shallow coastal waters of Panthalassa were particularly rich in invertebrate species. Trilobites, one of the most iconic groups of marine arthropods, were abundant during the early Paleozoic, though they eventually declined towards the late Permian period. These creatures, with their segmented bodies and exoskeletons, played a critical role in the marine food chain as both predators and prey.

Ammonites, a type of cephalopod, were another hallmark species of Panthalassa's biodiversity. These spiral-shelled creatures were closely related to modern squids and octopuses and ranged from small, delicate species to large, predatory forms. Ammonites were highly adaptable and thrived in various marine environments, including shallow seas and deeper waters. Their shells are now commonly found as fossils and serve as key indicators of the Mesozoic marine environment.

Brachiopods, another group of marine invertebrates that resembled clams, were common in Panthalassa. These filter-feeding organisms were particularly abundant in shallow waters, where they formed dense colonies on the seafloor. Though brachiopods look similar to modern bivalves (such as clams and mussels), they belong to a different evolutionary lineage and were far more diverse in ancient oceans.

Crustaceans, including ancient forms of shrimp, lobsters, and crabs, were also well represented in Panthalassa's ecosystems. These creatures played crucial roles as both scavengers and prey for larger predators, contributing to the complexity of the marine food web.

Coral-like Organisms and Early Reefs

While modern coral reefs are widespread and highly complex ecosystems, the coral reefs of Panthalassa were simpler and less diverse. During the Paleozoic, ancient corals such as rugose and tabulate corals built small reefs along Panthalassa's shallow coasts. These early reefs were home to a variety of marine species, including mollusks, echinoderms, and other filter-feeding organisms.

Deeper Ocean Life

The deeper regions of Panthalassa were home to more primitive life forms, as the conditions in the abyssal zones were harsher and less conducive to high biodiversity. In these deep-water environments, organisms had to adapt to cold temperatures, high pressure, and limited light. These conditions favored the evolution of scavengers, bottom-dwelling organisms, and creatures that could survive on the scarce nutrients sinking from the surface.

4.2 Marine Reptiles and Predators

One of the most remarkable aspects of Panthalassa's marine life was the emergence of gigantic marine reptiles that dominated the ocean's food chains during the Mesozoic era. These reptiles evolved from land-dwelling ancestors and adapted to life in the ocean, becoming some of the top predators of their time.

Ichthyosaurs: The "Fish Lizards"

Ichthyosaurs, meaning "fish lizards," were among the most iconic marine reptiles that swam the waters of Panthalassa. These creatures first appeared during the early Triassic period and quickly became dominant predators. Ichthyosaurs had streamlined bodies similar to modern dolphins or sharks, with large eyes and long snouts filled with sharp teeth, perfect for catching fish and squid.

Some ichthyosaurs, such as *Shonisaurus*, grew to enormous sizes, reaching lengths of up to 15 meters (50 feet). These massive reptiles likely hunted large prey, including cephalopods and other marine reptiles. Their powerful tails allowed them to swim at high speeds, making them effective hunters in the open ocean.

Plesiosaurs: Long-Necked Predators

Another group of marine reptiles that flourished in Panthalassa was the plesiosaurs. These creatures are best known for their long necks and broad, paddle-like limbs, which allowed them to glide gracefully through the water. Plesiosaurs had relatively small heads compared to their long necks, and they used their flexible necks to ambush prey such as fish and smaller marine reptiles.

Some species of plesiosaurs, like *Elasmosaurus*, had necks that could stretch over seven meters (23 feet) in length, while others, such as pliosaurs, had shorter necks but much larger heads and powerful jaws, making them fearsome predators.

Mosasaurs: Apex Predators of the Late Cretaceous

In the later stages of Panthalassa's existence, particularly during the Late Cretaceous period, mosasaurs became the dominant marine predators. These large, serpentine reptiles could grow up to 18 meters (60 feet) in length and were top predators of the time. Mosasaurs had robust, streamlined bodies, powerful tails for swimming, and large, conical teeth designed for grasping prey.

Mosasaurs fed on a wide variety of marine life, including fish, ammonites, and even other marine reptiles. Some species may have ventured into shallower coastal waters to hunt, while others dominated the deeper regions of the ocean.

4.3 Coral Reefs and Shallow Ecosystems

Though Panthalassa was a massive ocean, its shallow coastal regions were biologically diverse, providing rich habitats for various marine organisms. Coral-like structures and smaller, less complex reefs played an important role in supporting marine life.

Early Coral Reefs

Panthalassa's coral reefs were not as extensive or diverse as modern reefs, but they were still important ecological hubs. These reefs were primarily constructed by ancient corals, such as rugose and tabulate corals, and were much smaller and less intricate than today's coral systems. These reef structures provided habitats for a wide range of organisms, including bivalves, sponges, and early crustaceans.

Bivalves and Gastropods

Shallow marine environments in Panthalassa were home to a variety of bivalves and gastropods. Bivalves, such as clams and oysters, anchored themselves to the seafloor or to rocks in reef environments, filtering nutrients from the water. Gastropods, which include snails, were abundant in these habitats and played crucial roles in the ecosystem as both herbivores and scavengers.

Echinoderms: Sea Stars and Crinoids

Echinoderms, such as sea stars, brittle stars, and crinoids (often called "sea lilies"), were common in the shallow waters of Panthalassa. These creatures were filter feeders, scavengers, and predators, and they thrived in reef ecosystems and on soft seafloors. Crinoids, with their long, feather-like arms, were particularly common during the Paleozoic and Mesozoic eras, forming large colonies in shallow waters.

Early Marine Ecosystems

In addition to coral reefs, shallow ecosystems in Panthalassa included seagrass beds, kelp forests, and other marine plant life. These ecosystems supported herbivorous fish, marine invertebrates, and smaller reptiles, creating a diverse web of life. The relatively warm, shallow waters along Panthalassa's continental margins were ideal for fostering these vibrant marine communities.

Chapter 5: The Legacy of Panthalassa

Panthalassa, the massive superocean that surrounded the supercontinent Pangaea, left an enduring legacy that continues to shape Earth's oceans, geography, and climate today. This chapter delves into how Panthalassa evolved into the modern Pacific Ocean, what geological evidence remains from this ancient body of water, and the broader role Panthalassa played in Earth's dynamic history.

5.1 The Formation of the Pacific Ocean

The eventual fragmentation of Panthalassa began during the Jurassic period, approximately 180 million years ago, as the supercontinent Pangaea started to break apart due to tectonic forces. The splitting of Pangaea created the Atlantic and Indian Oceans, while Panthalassa, which had been the dominant ocean for hundreds of millions of years, slowly transformed into the Pacific Ocean. This transformation marked a significant turning point in Earth's geological history, with Panthalassa evolving into the world's largest modern ocean, the Pacific.

Tectonic Forces and the Breakup of Panthalassa

The breakup of Panthalassa was driven by the same tectonic forces that led to the disintegration of Pangaea. The Earth's lithosphere, composed of rigid tectonic plates, was constantly in motion due to convection currents in the mantle. These currents created divergence zones, where plates moved apart, and convergence zones, where plates collided and subducted beneath one another. As Pangaea split into the continents of Laurasia in the north and Gondwana in the south, Panthalassa's vast waters were divided into multiple oceanic basins.

At the center of this transformation was the creation of the Pacific Ocean, which retained many of the features that once characterized Panthalassa. As new oceanic crust formed at mid-ocean ridges, the Pacific expanded, while older portions of Panthalassa's ocean floor subducted under continental plates, giving rise to some of the deepest oceanic trenches and volcanic island arcs that we see today.

Features of the Pacific Ocean from Panthalassa

The Pacific Ocean, as the direct descendant of Panthalassa, inherited many of its geological features, particularly its vast expanse, deep ocean trenches, and volcanic island chains. The Pacific, covering more than 63 million square miles, remains the largest ocean on Earth, much like

Panthalassa dominated the Earth's surface during the Paleozoic and Mesozoic eras.

- **Deep Trenches**: One of the most notable features inherited from Panthalassa are the deep oceanic trenches, such as the Mariana Trench, the deepest part of the Earth's oceans, reaching depths of over 36,000 feet. These trenches formed due to subduction zones where older sections of the oceanic crust were forced beneath continental plates—a process that began in Panthalassa and continues today in the Pacific.
- **Island Arcs**: The Pacific is also home to numerous volcanic island arcs, such as the Aleutian Islands, the Japanese Archipelago, and the Philippines, formed from subduction-related volcanic activity. These island chains trace their origins back to the tectonic movements and volcanic activity in the Panthalassa Ocean. As the Pacific plate subducted beneath adjacent plates, it created molten rock that rose to the surface, forming these island arcs.
- **Mid-Ocean Ridges**: The Pacific Ocean floor is also shaped by mid-ocean ridges, where new oceanic crust is formed. These ridges, remnants of Panthalassa's seafloor-spreading activity, continue to add new material to the ocean floor, expanding the Pacific in a manner similar to how Panthalassa once grew.

Climatic and Oceanic Influence

The Pacific Ocean, like Panthalassa, continues to have a profound influence on Earth's climate and weather patterns. The vast size of the Pacific means it plays a critical role in the global distribution of heat and moisture, influencing weather systems, ocean currents, and marine ecosystems. Oceanic phenomena such as El Niño and La Niña, which have far-reaching effects on global weather patterns, are direct results of atmospheric and oceanic interactions within the Pacific. These climate-shaping forces have their origins in the ancient dynamics of Panthalassa.

5.2 Modern Geological Evidence

Understanding the history of Panthalassa and its transition into the Pacific Ocean relies on a combination of geological and paleontological evidence. Geologists and paleontologists study remnants of the ancient oceanic crust, magnetic patterns on the ocean floor, and fossil records to reconstruct the formation and evolution of Panthalassa.

Ocean Floor Sediments and Plate Tectonics

The ocean floor preserves a detailed geological record of Earth's history, including the existence of Panthalassa. Modern geologists study ocean floor sediments to piece together the puzzle of Panthalassa's evolution. As tectonic plates moved and spread apart, new oceanic crust was continuously generated at mid-ocean ridges. The layers of sediments deposited on the ocean floor provide insights into the changes in marine environments, shifts in global climate, and the processes of plate tectonics that shaped Panthalassa.

- **Sediment Analysis**: Sediments on the ocean floor, including those found in deep-sea cores, offer a glimpse into Panthalassa's history. By examining the composition of these sediments—such as carbonate shells from marine organisms—scientists can determine past ocean temperatures, salinity, and levels of oxygen. These records reveal how Panthalassa supported marine ecosystems and how its conditions changed during periods of mass extinction, such as the Permian-Triassic event.

- **Subduction and Preservation of Ancient Crust**: As Panthalassa's oceanic crust was subducted beneath continental plates, much of its ancient seafloor was lost to Earth's interior. However, some remnants of Panthalassa's crust have been preserved in today's Pacific Ocean, especially in regions where subduction processes were incomplete or slower. Geologists use these preserved sections of oceanic crust to study the movement of tectonic plates over millions of years.

Magnetic Patterns and Fossil Evidence

Another key piece of evidence comes from the study of magnetic patterns in oceanic crust. As new crust forms at mid-ocean ridges, iron-rich minerals in the molten rock align with Earth's magnetic field, preserving a "snapshot" of the magnetic polarity at the time of formation. These magnetic stripes provide a timeline of ocean floor creation and movement, allowing scientists to track the age and expansion of ancient oceanic crust.

- **Magnetic Stripes**: In Panthalassa, as in today's oceans, magnetic minerals recorded reversals of Earth's magnetic field, creating alternating stripes of normal and reversed polarity on the ocean floor. By mapping these magnetic patterns in the Pacific, geologists can trace the movement of tectonic plates and the history of Panthalassa's oceanic spreading centers.
- **Fossil Evidence**: Fossil records from marine sediments also offer crucial insights into the ecosystems that thrived in Panthalassa. Fossils of ammonites, trilobites, and other marine organisms help scientists understand the biodiversity of Panthalassa and how it was impacted by major extinction events, such as the Great Permian Extinction.

5.3 Panthalassa's Role in Earth's History

Panthalassa was not just a vast body of water; it played an essential role in shaping Earth's history, influencing climate, marine life, and the movement of continents.

Climate Regulation

As the largest ocean in Earth's history, Panthalassa acted as a massive heat sink, regulating global temperatures and influencing climate patterns across the planet. Ocean currents within Panthalassa redistributed heat between the equator and the poles, moderating the Earth's climate and creating stable conditions for the evolution of life.

Panthalassa's sheer size helped to stabilize atmospheric conditions, preventing extreme temperature fluctuations and allowing the development of complex ecosystems on land and in the ocean. This climate regulation was particularly important during the late Paleozoic and early Mesozoic, when Earth's atmosphere and ecosystems were in a delicate balance.

Support for Marine Life

Panthalassa was home to a diverse array of marine organisms, from ancient fish and invertebrates to massive marine reptiles. The ocean supported vast ecosystems, with shallow continental shelves providing habitats for coral reefs and deeper waters offering environments for a range of marine predators. The biodiversity of Panthalassa laid the foundation for the evolution of modern marine life, with many of its species evolving into the ancestors of today's ocean dwellers.

Influence on Continental Drift

Panthalassa was intimately connected to the process of continental drift, as the movement of Earth's tectonic plates over millions of years reshaped the planet's surface. The superocean played a key role in the cycle of plate tectonics, with subduction zones around its perimeter pulling plates beneath continental margins and driving the formation of mountain ranges, island arcs, and oceanic trenches. This process not only shaped the geography of Earth but also influenced global climate, sea levels, and the distribution of life.

Mass Extinction Events

Panthalassa was also witness to some of the most dramatic events in Earth's history, including the Permian-Triassic extinction event, which wiped out the majority of marine and terrestrial species. The environmental changes associated with this mass extinction, including ocean acidification and anoxic conditions in Panthalassa's deep waters, had profound effects on marine life and the course of evolutionary history. These extinction events shaped the future of life on Earth and set the stage for the rise of the dinosaurs in the Mesozoic.

In summary, Panthalassa's legacy is evident in the modern Pacific Ocean and in the geological processes that continue to shape our planet. The study of Panthalassa offers valuable insights into the ancient Earth, from its role in regulating the climate to its influence on tectonic activity and marine ecosystems. Through the lens of Panthalassa, we gain a deeper understanding of the dynamic and interconnected forces that have shaped our world over billions of years.

Chapter 6: Conclusion – The Ocean That Shaped a World

Panthalassa was more than just a vast, ancient body of water. It was a central player in Earth's dynamic and transformative geological history, as well as an essential contributor to the biological evolution that unfolded across the planet's oceans and landmasses. The significance of Panthalassa extends far beyond its physical dimensions. Its presence directly influenced the formation, stability, and eventual fragmentation of the supercontinent Pangaea, as well as the environmental conditions that would govern much of Earth's climate and biosphere for hundreds of millions of years.

6.1 The Role of Panthalassa in Earth's Geological Evolution

Panthalassa was integrally linked to the forces that shaped the Earth's surface. As the supercontinent Pangaea formed about 300 million years ago, the entirety of Earth's landmass was consolidated into a single giant continent, and Panthalassa surrounded it as the dominant ocean. This arrangement was a product of plate tectonics, the geological processes that cause the movement of the Earth's lithospheric plates. These processes, including subduction (where oceanic crust sinks into the mantle) and seafloor spreading, drove the continual evolution of Panthalassa's shape and size.

The vastness of Panthalassa also created a stable, largely uninterrupted oceanic environment that spanned from the North Pole to the South Pole. This colossal body of water contributed to relatively consistent climatic conditions across large swaths of the Earth. Its currents played a crucial role in regulating global temperatures, transporting heat from the equator to the poles and shaping the Earth's weather patterns. As such, Panthalassa acted as a planetary moderator, helping to stabilize Earth's climate.

The oceanic crust beneath Panthalassa was constantly being formed at mid-ocean ridges and destroyed at subduction zones along its perimeter. The forces at work within Panthalassa's depths were responsible for the development of volcanic island arcs and deep oceanic trenches. For instance, the proto-Ring of Fire, which would later define the Pacific Ocean's seismic and volcanic activity, had its origins in the tectonic processes that operated beneath Panthalassa.

6.2 Biological Importance of Panthalassa

Panthalassa was not only a geological powerhouse but also a thriving cradle of life. Its ecosystems were home to an extraordinary diversity of marine organisms, ranging from simple microorganisms to complex marine reptiles. The life that inhabited Panthalassa contributed to the biodiversity of Earth's early oceans, influencing the evolutionary pathways of countless species.

Marine organisms, including ammonites, trilobites, brachiopods, and early fish, flourished in the shallow coastal areas and the vast open waters of Panthalassa. These species played crucial roles in marine food webs and served as early indicators of environmental changes. For example, during periods of oceanic anoxia (low oxygen levels), many of these marine species experienced significant declines, signaling shifts in ocean chemistry that were often associated with broader climatic or geological upheavals.

Perhaps most notably, Panthalassa's waters were witness to some of the most significant extinction events in Earth's history, including the Permian-Triassic extinction event, also known as "The Great Dying." This catastrophic event, which occurred approximately 252 million years ago, wiped out more than 90% of marine species and drastically altered the planet's biosphere. Volcanic eruptions, climate change, and disruptions in ocean circulation may have caused oxygen levels in Panthalassa to plummet, leading to mass die-offs of marine organisms. This event underscores Panthalassa's role not only as a habitat for life but also as a dynamic and at times destructive force in Earth's history.

6.3 The Legacy of Panthalassa

Though Panthalassa no longer exists as a single unified ocean, its legacy lives on in the Pacific Ocean, which is a direct descendant of the Superocean. After the breakup of Pangaea during the Jurassic period, Panthalassa began to fragment, giving rise to the Atlantic, Indian, and Southern Oceans while the vast Pacific remained as the largest vestige of the Superocean. Today, the Pacific Ocean still occupies a dominant position on Earth's surface, covering over 63 million square miles, and it remains home to some of the planet's most significant tectonic and oceanic features, such as the Mariana Trench, the world's deepest point.

The geological processes that shaped Panthalassa, such as subduction and seafloor spreading, continue to drive the evolution of the Pacific Ocean today. The Ring of Fire, a horseshoe-shaped zone of intense volcanic and seismic activity encircling the Pacific, is a direct result of the tectonic forces that once operated beneath Panthalassa. This ongoing activity serves as a reminder of the dynamic forces that once governed the Superocean.

Panthalassa also left an imprint on the continents that surrounded it. The ancient oceanic crust that formed the floor of Panthalassa has long since been subducted into the Earth's mantle, but remnants of this crust can still be found in certain geological formations. These remnants help geologists and paleontologists reconstruct the history of Panthalassa and gain a deeper understanding of the processes that shaped Earth's surface millions of years ago.

6.4 Future Oceans and Continental Drift

The disappearance of Panthalassa was not the end of the forces that created it. Continental drift, driven by plate tectonics, continues to reshape the Earth's surface. The continents are still moving today, albeit at a pace of a few centimeters per year. These movements are part of a larger, ongoing cycle that will likely lead to the formation of a new supercontinent in the distant future—a process known as the "supercontinent cycle."

As the Pacific Ocean continues to shrink due to subduction, the Atlantic Ocean is widening, creating new oceanic crust at the Mid-Atlantic Ridge. Scientists predict that, millions of years from now, the continents will once again collide and merge, forming a new supercontinent— sometimes referred to as "Pangaea Proxima" or "Amasia." This new supercontinent will be surrounded by a new ocean, potentially similar in scale to Panthalassa.

The formation of this future supercontinent and its surrounding ocean will bring about significant changes in Earth's climate, geography, and biological ecosystems. Just as Panthalassa helped regulate Earth's climate during the time of Pangaea, this new ocean will influence future climate patterns, potentially altering ocean currents, wind systems, and global temperatures.

While the specifics of these future changes are uncertain, the continued movement of Earth's tectonic plates guarantees that the cycle that began with Panthalassa is far from over. The processes that governed the Superocean's creation, stability, and eventual demise are still at work beneath the Earth's surface, ensuring that the planet's oceans and continents will continue to evolve over geological timescales.

6.5 Conclusion: The Enduring Influence of Panthalassa

Panthalassa was not just a passive body of water; it was an active and integral component of Earth's geological and biological history. It shaped the climate, influenced the movement of continents, and supported diverse ecosystems that contributed to the planet's evolutionary history. Its eventual transformation into the Pacific Ocean and the fragmentation of Pangaea marked the end of an era in Earth's history, but the forces that created and sustained Panthalassa are still at work today.

The study of Panthalassa provides a window into Earth's deep past, offering insights into the processes that have shaped our planet for billions of years. By understanding the history of Panthalassa, we gain a greater appreciation for the dynamic nature of Earth's oceans and their ongoing influence on the planet's geological and environmental systems. Even though Panthalassa no longer exists as a singular ocean, its legacy endures in the Pacific, in the geological formations it left behind, and in the continued movement of Earth's tectonic plates.

As we look to the future, the forces that once governed Panthalassa will continue to shape Earth's surface, ensuring that the planet's oceans, continents, and ecosystems will remain in constant flux. Just as Panthalassa was the ocean that shaped a world, the oceans of the future will play a similarly crucial role in determining the fate of Earth's continents and lifeforms.

Bibliography

1. **"Oceans: A Very Short Introduction"** by Dorrik Stow

A concise introduction to oceanography, covering the origins, structure, and significance of Earth's oceans, including ancient superoceans like Panthalassa.

2. **"Earth's Deep History: How It Was Discovered and Why It Matters"** by Martin J. S. Rudwick

A detailed exploration of the discovery of Earth's geological history, focusing on how scientists pieced together the planet's ancient past, including the existence of superoceans.

3. **"The Ocean of Life: The Fate of Man and the Sea"** by Callum Roberts

This book looks at the deep connection between oceans and life on Earth, providing context for how oceans like Panthalassa supported ancient marine ecosystems.

4. **"The Ends of the World: Volcanic Apocalypses, Lethal Oceans, and Our Quest to Understand Earth's Past Mass Extinctions"** by Peter Brannen

This book examines mass extinction events, including those that occurred during the time Panthalassa existed, such as the Permian-Triassic extinction.

5. "The Continental Drift Controversy: Wegener and the Early Debate" by Henry R. Frankel

A historical account of the development of continental drift theory, which is crucial to understanding how superoceans like Panthalassa formed and evolved.

6. "Sea Change: A Message of the Oceans" by Sylvia A. Earle

Renowned oceanographer Sylvia Earle discusses the importance of oceans, both ancient and modern, and their role in supporting life, with insights that could be applied to Panthalassa.

7. "Earth's Evolving Systems: The History of Planet Earth" by Ronald E. Martin

A textbook that covers the geological and biological history of Earth, with specific sections on the Paleozoic and Mesozoic eras when Panthalassa existed.

8. "The Sea Around Us" by Rachel Carson

While primarily focused on modern oceans, Carson's lyrical writing about the ocean's history and its connection to life can offer inspiration for discussing Panthalassa's significance.

9. "Trilobite! Eyewitness to Evolution" by Richard Fortey

A fascinating exploration of trilobites, ancient marine creatures that lived in oceans like Panthalassa, providing insight into the life that inhabited the superocean.

Acknowledgments

Writing *Panthalassa: The Superocean – The Vast Body that Shaped the Ancient Earth* has been a remarkable journey, and it would not have been possible without the support and contributions of many individuals.

First and foremost, I would like to express my deepest gratitude to my family and friends for their constant encouragement and unwavering belief in me throughout this project. Your patience, understanding, and support provided the foundation I needed to complete this book.

I am also deeply thankful to the many researchers, geologists, and oceanographers whose groundbreaking work on tectonics, paleoclimatology, and marine biology has enriched our understanding of Earth's history and the ancient oceans. Their work has served as the backbone of the knowledge presented in this book. A special thank you to those whose published works and academic resources helped illuminate the fascinating world of Panthalassa.

To my editor and proofreader, your keen eyes and thoughtful suggestions have been invaluable in refining and shaping the final manuscript. Your attention to detail and dedication to ensuring clarity and accuracy has helped me present complex scientific concepts in an accessible and engaging way.

I would also like to extend my thanks to my publisher and the entire team behind the scenes who have supported this book from its inception to its publication. Your commitment to bringing this work to life has been truly inspiring.

Finally, I would like to thank you, the readers, for your curiosity and interest in the incredible story of Panthalassa and the ancient Earth. It is my hope that this book sparks your imagination and deepens your appreciation for the forces that have shaped our planet over billions of years.

To everyone who has supported me on this journey—thank you.

With heartfelt appreciation,

Zahid Ameer
Versatile Indie Author

Disclaimer

The information presented in *Panthalassa: The Superocean - The Vast Body that Shaped the Ancient Earth* is intended for educational and informational purposes only. While every effort has been made to ensure the accuracy of the content, the details and interpretations provided are based on current scientific understanding, which may evolve as new research and discoveries emerge. The author and publisher do not claim to provide definitive conclusions or exhaustive coverage on the subject. Readers are encouraged to consult primary scientific sources and experts in the field for further study.

The historical, geological, and scientific information contained in this book is provided "as is" and is not intended to serve as professional advice. The author and publisher disclaim any responsibility or liability for any actions or consequences arising from the use of this information.

By reading this book, you acknowledge and accept that the understanding of Earth's ancient oceans and supercontinents is constantly evolving, and that this work may not reflect the most recent developments in geology, paleontology, or related disciplines.

About me

I am Zahid Ameer, hailing from the vibrant country of India. As an author, ghostwriter, bibliophile, online affiliate marketer, blogger, YouTuber, graphic designer, and animal lover, I have woven my passions into a unique tapestry that defines my life's work.

Born and raised in India, I have always possessed a deep love for literature. With an insatiable appetite for books, I have amassed an impressive collection of around 1,600 titles, predominantly in English. My passion for reading brings me immense joy and serves as a source of inspiration for my writing endeavors.

I have compiled an impressive portfolio of written works as an author and ghostwriter. With a captivating writing style and an innate ability to craft engaging narratives, I bring my stories to life, captivating readers from all walks of life. My wide range of interests and experiences contribute to the richness of my writing, allowing me to connect with my audience on a heartfelt level effortlessly.

Beyond my literary pursuits, I have also established a strong presence on various digital platforms. I utilize my YouTube channel and blog to raise awareness about all types of knowledge and to share heartwarming stories of animals. Using my platform to shed light on important issues, I strive to create a world where humans and animals can coexist harmoniously.

In addition to my work as an author, I have also dabbled in the world of affiliate marketing. With my webpreneur spirit, I have ventured into online marketing, leveraging my knowledge and skills to promote products and services that align with my values.

However, my most cherished role is that of a father. Family is at the core of my being, and everything I do is centered around creating a better future for my loved ones. My dedication to my family is evident in my passion for personal growth and my relentless pursuit of success. Through my various endeavors, I strive to set an example of perseverance and ambition for my children, inspiring them to chase their dreams unapologetically.

In a world where specialization often dominates, I defy convention by embracing multiple passions and excelling in diverse fields. My love for books, animals, and family has become the driving force behind my achievements. By the grace of Almighty God, my unique blend of characteristics has allowed me to leave an indelible mark on the world, enriching the lives of those I encounter along the way.

To your grand success in life,

Zahid Ameer
Versatile Indie Author

www.ingramcontent.com/pod-product-compliance
Lightning Source LLC
Chambersburg PA
CBHW070130230526
45472CB00004B/1502